U0309428

上海高度

Shanghai Height

姜庆共 著

Jiang Qinggong

同济大学出版社·上海
TONGJI UNIVERSITY PRESS SHANGHAI

“城市行走”书系
策划：江岱、姜庆共

责任编辑：江岱
助理编辑：罗璇
图文：姜庆共
设计助理：周祺
翻译：许昊旸

鸣谢：张亚东、李彦伯、
王伟强、杨辰、刘鹏、
上海自来水科技馆

Citywalk Series
Curator: Jiang Dai, Jiang Qinggong

Editor: Jiang Dai
Assistant Editor: Luo Xuan
Illustrater & Text: Jiang Qinggong
Assistant Designer : Zhou Qi
Translator: Xu Haoyang

Acknowledgements: Zhang Yadong,
Li Yanbo, Wang Weiqiang,
Yang Chen, Liu Peng,
Shanghai Waterworks Science and
Technology Museum

上海高度

姜庆共 著

Shanghai Height

Jiang Qinggong

同济大学出版社
TONGJI UNIVERSITY PRESS

上海境内自然地貌平坦。今陆地最高处为松江区的天马山，海拔 98.2 米。位于金山区的大金山岛，海拔 103.4 米，是上海的最高峰

The vast majority of Shanghai's land area is flat. On land the highest point is Tianma Hill (98.2 m above sea level), in Songjiang District. At 103.4 m above sea level, the peak of Dajinshan Island, in Jinshan Disctrict, is the highest point within the Shanghai municipality.

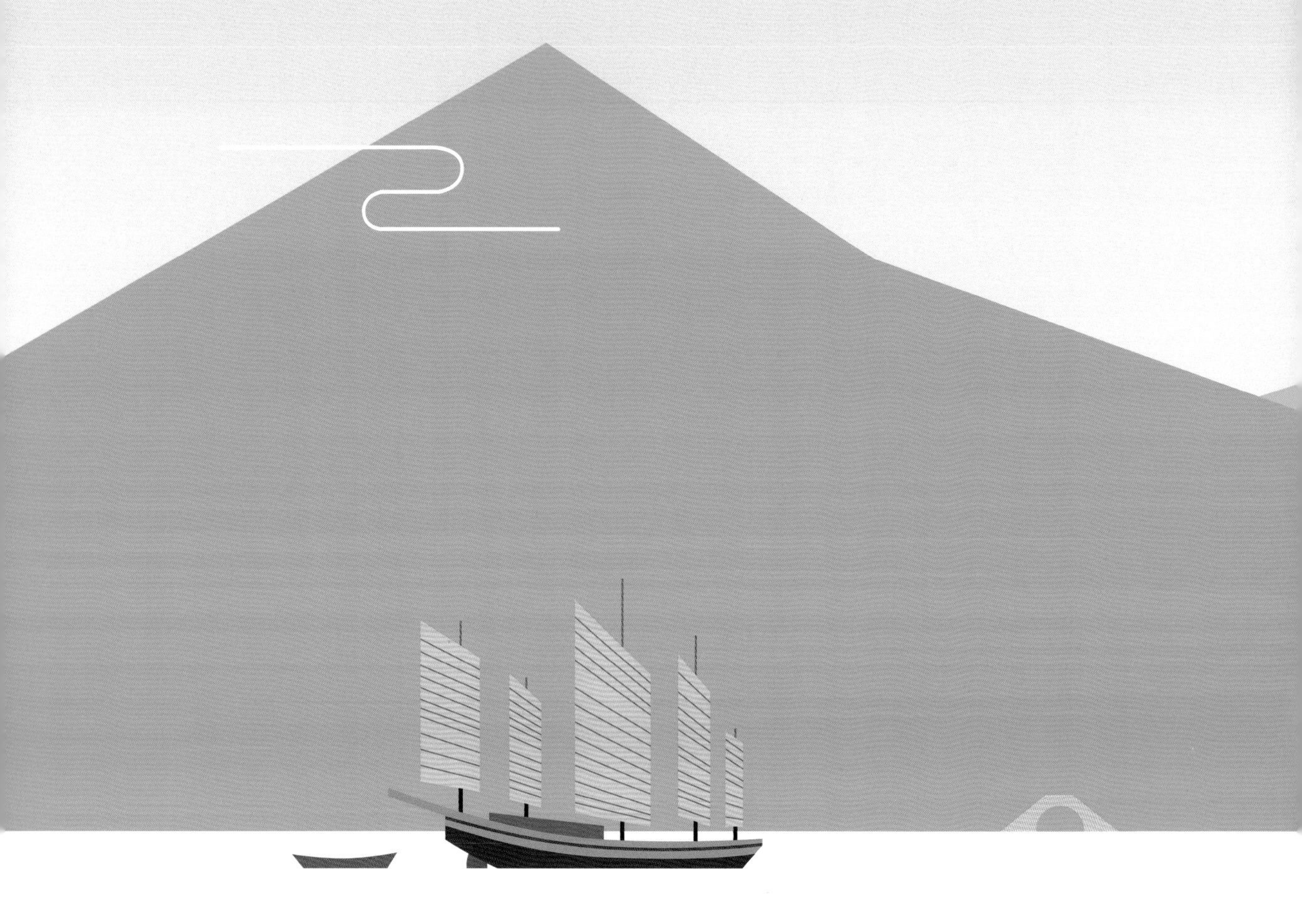

建于唐朝的松江唐经幢，是上海现存最古老的石刻建筑，残高 9.3 米

The Dharani Column Building of the Tang Dynasty in Songjiang, with a remaining height of 9.3 m, is the oldest stone carved architecture in Shanghai.

青龙塔位于唐宋时期江南地区对外贸易繁盛的青龙镇故址，初建于唐朝，北宋重建，现残高约 30 米

The Qinglong Pagoda (remaining height: about 30 m), built in the Tang Dynasty and rebuilt in the N. Song Dynasty, lies at the former site of Qinglong Town, which witnessed the booming foreign trade in the Jiangnan region during Tang and Song.

龙华塔相传初建于三国，北宋重建，高 40.64 米，曾为象征上海的主要地标之一

The Longhua Pagoda (40.64 m), said to be first erected in the period of the Three Kingdoms and rebuilt in the N. Song Dynasty, was one of the major landmarks of Shanghai.

明朝时期上海县围筑城墙，高近 8 米，现存 50 余米长旧址及大境阁

The city wall of the Shanghai County during Ming Dynasty, nearly 8 m in height, now remains about 50 m long with the Dajing Ge Pavilion.

石库门里弄是 20 世纪上海最具代表性的民居建筑，吉祥里为现存最早的石库门里弄之一，建于 1876 年以前，高 10 ~ 11 米

Shikumen Lilong represents Shanghai's residential architecture of the 20th century. Jixiang Li, built before 1876, was one of the earliest Lilong that survived till today. Its height ranges from 10 m to 11 m.

1883 年中国第一座现代化水厂杨树浦水厂在上海建成，现存烟囱建于 1928 年，高约 29 米

Yangshupu Water Works, the first modern Chinese water supply and purification plant, was established in Shanghai in 1883. The remaining chimney was built in 1928 and is about 29 m high.

外滩气象信号台是上海最早的气象预报信号台，初建于 1884 年，1908 年重建，高 49.8 米

The Gutzlaff Signal Tower was originally erected in 1884, as the first signal tower in Shanghai to provide weather information, in particular typhoon warnings. It was re-built in 1908 and 49.8 m high.

徐家汇天主堂建于 1910 年，高 56.6 米，是上海最大的教堂

St. Ignatius Cathedral, built in 1910 and 56.6 m in height, is the largest cathedral in Shanghai.

1917 年先施公司建成上海第一家由华人创办的大型环球百货商场，高近 60 米，并最早开设屋顶游乐场，聘用女店员，实行商品明码不二价

In 1917, Sincere Department Store (nearly 60 m in height), the first global department store founded by Chinese, was open in Shanghai. It was also the first to have amusement facilities on the roof terrace, to hire female shop assistants and to set labelled prices.

《字林西报》是上海最早出版的英文报纸之一，1924 年在外滩新址建成字林大楼，高 40.2 米，是当年外滩最高的建筑

North China Daily News was one of the earliest English newspapers published in Shanghai. In 1924, the North China Daily News Building (40.2 m) was built, then the tallest building on the Bund.

大世界游乐场曾为闻名全国的游乐场，原二层建筑初建于 1917 年，1924 年改建并加建塔楼，高 55.3 米

The Great World was a nationwide well-known amusement arcade. The original two-storeyed building was built in 1917. It was reconstructed in 1924, topped with a tower and reaches 55.3 m.

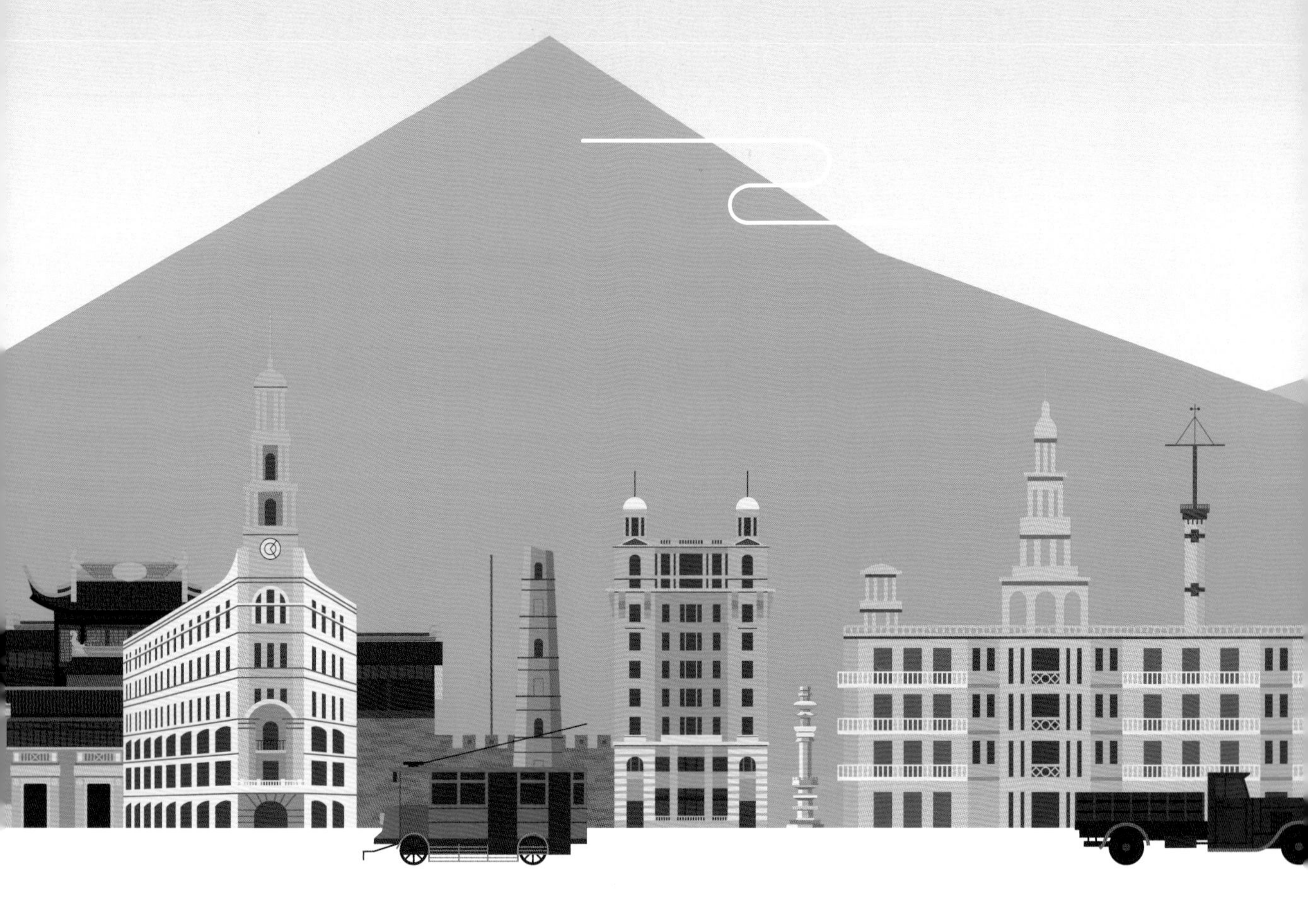

位于苏州河畔的上海邮政总局建于 1924 年，高 51.16 米，是中国最早的邮政大楼

The General Post Office of Shanghai (51.6 m), built in 1924 on the Suzhou Creek, was China's earliest post office building.

江海关大楼建于 1927 年，高 70 米，大楼顶部为钟楼，报时一刻，钟声缭绕浦江上空

The Customs House (70 m), built in 1927, stands with a clock tower atop. The music, when the clock is striking, resonates across the Huangpu River.

外滩历史建筑群中最高的建筑为沙逊大厦，建于 1928 年，高 77 米

The Sassoon House (77 m), completed in 1928, is the highest historical building of the Bund.

国际饭店建于 1934 年，其高度 83.8 米，保持了上海大楼的最高纪录近 50 年

Park Hotel (83.8 m), built in 1934, held the record of the highest building in Shanghai for almost 50 years.

百老汇大厦建于 1934 年，高 76.7 米，与外白渡桥相临，曾为上海著名的地标建筑

The Broadway Mansions (76.7 m), built in 1934 and at the northern end of the Waibaidu Bridge (Garden Bridge), was Shanghai's popular landmark.

上海市立图书馆建于 1935 年，高 25 米，为上海特别市政府“大上海计划”工程之一

Shanghai Municipal Library (25 m), built in 1935, was one of the project of the Greater Shanghai Plan launched by the Shanghai Special City Government.

上海最早使用自动扶梯的商场大新公司，建于 1936 年，高 42.3 米，曾为闻名全国的百货商店

The Sun Department Store (42.3 m), built in 1936, was famous nationwide and the first department store to install escalators in Shanghai.

中国银行大楼是外滩历史建筑群中，唯一由华人设计师主持设计的大楼，建于 1937 年，高 69 米

Bank of China Building (69 m), built in 1937, was among the Bund historical architecture complex the only building designed by a Chinese in charge.

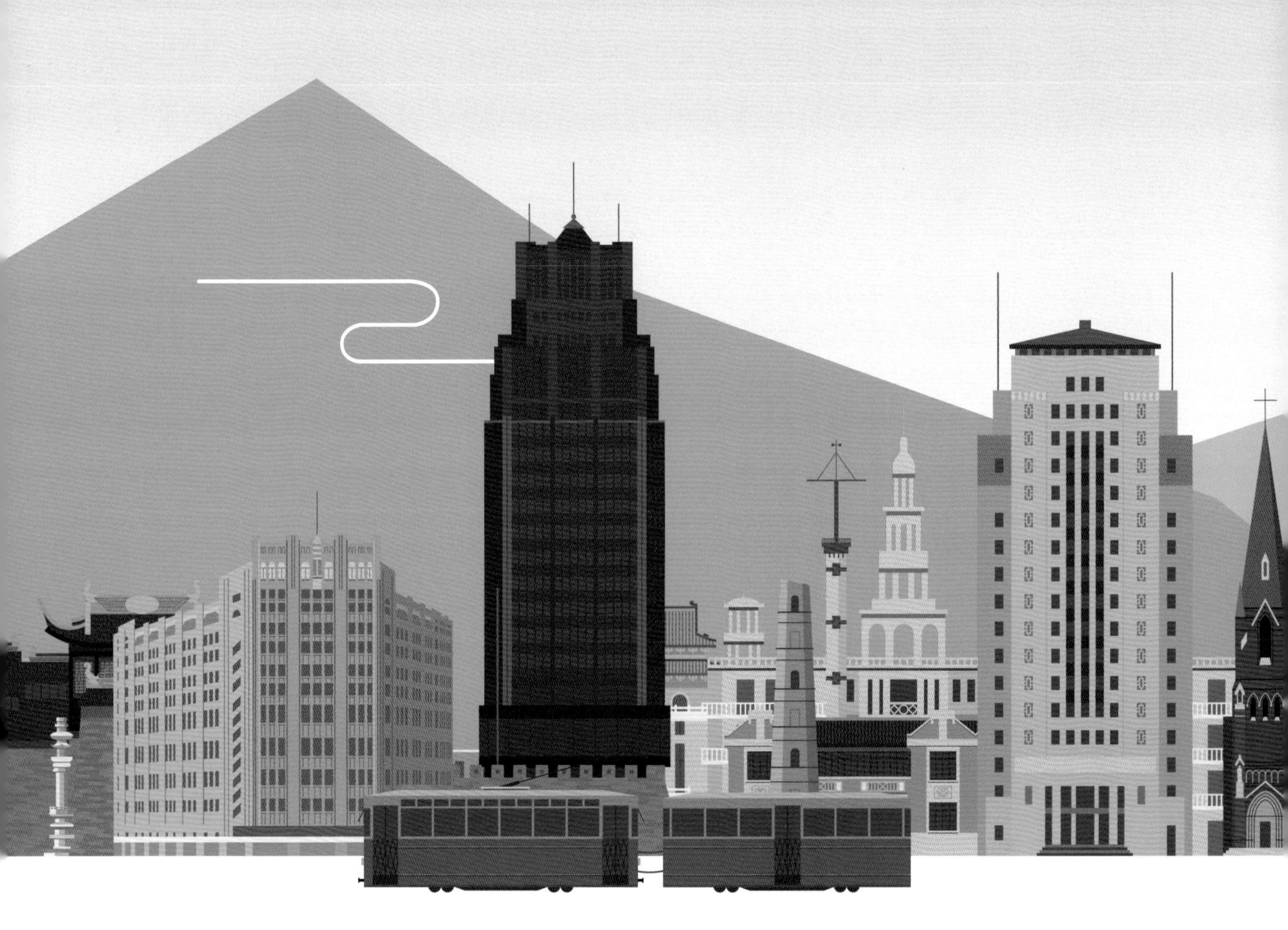

1952 年新中国第一个工人新村曹杨新村在上海建成，高 11 ～ 12 米

In 1952, New China's first residential district for the workers, Caoyang New Estate (11–12 m in height), was finished in Shanghai.

中苏友好大厦建于 1955 年，
其尖塔顶端高 110.4 米

Sino-Soviet Friendship Building,
whose sprire reaches 110.4 m,
was built in 1955.

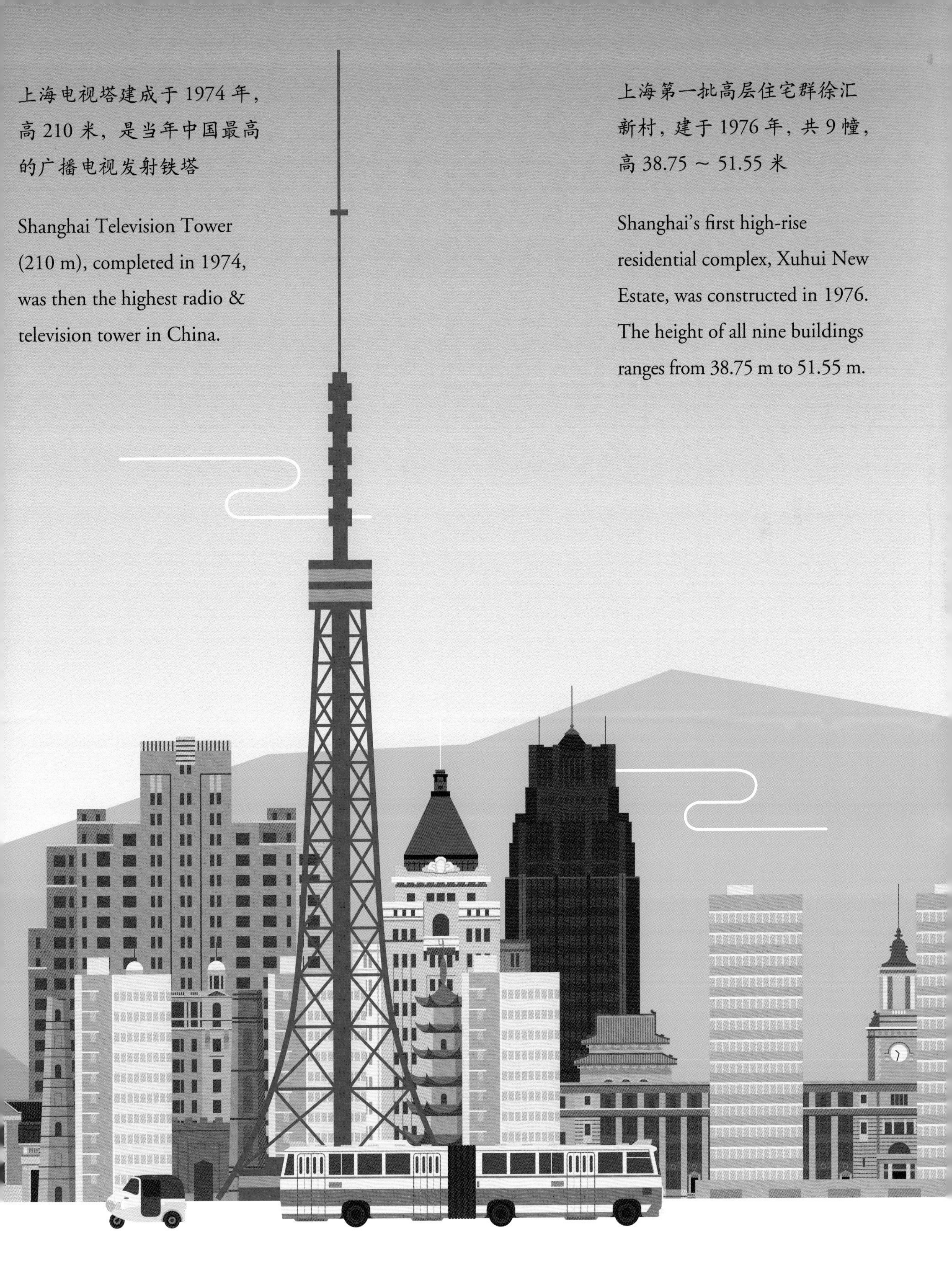

上海电视塔建成于 1974 年，高 210 米，是当年中国最高的广播电视发射铁塔

Shanghai Television Tower (210 m), completed in 1974, was then the highest radio & television tower in China.

上海第一批高层住宅群徐汇新村，建于 1976 年，共 9 幢，高 38.75 ~ 51.55 米

Shanghai's first high-rise residential complex, Xuhui New Estate, was constructed in 1976. The height of all nine buildings ranges from 38.75 m to 51.55 m.

为配合上海石油化工总厂建设，黄浦江上第一座公路、铁路桥梁黄浦江大桥于 1976 年建成，高 50 余米

In coordination with the construction of a petrochemical factory in Shanghai, the first road-rail bridge over the Huangpu River, Huangpujiang Bridge, was completed in 1976 with a height of over 50 m.

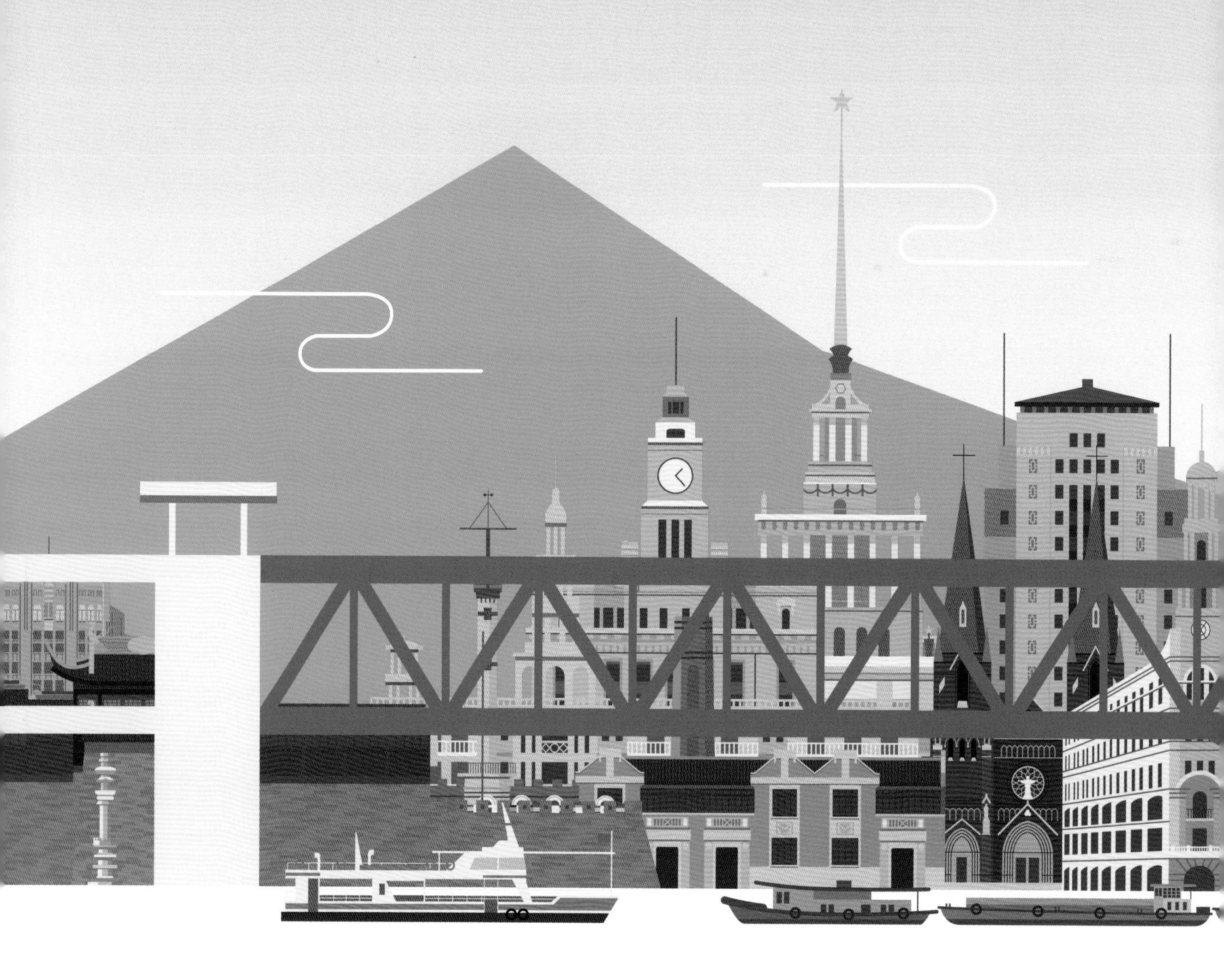

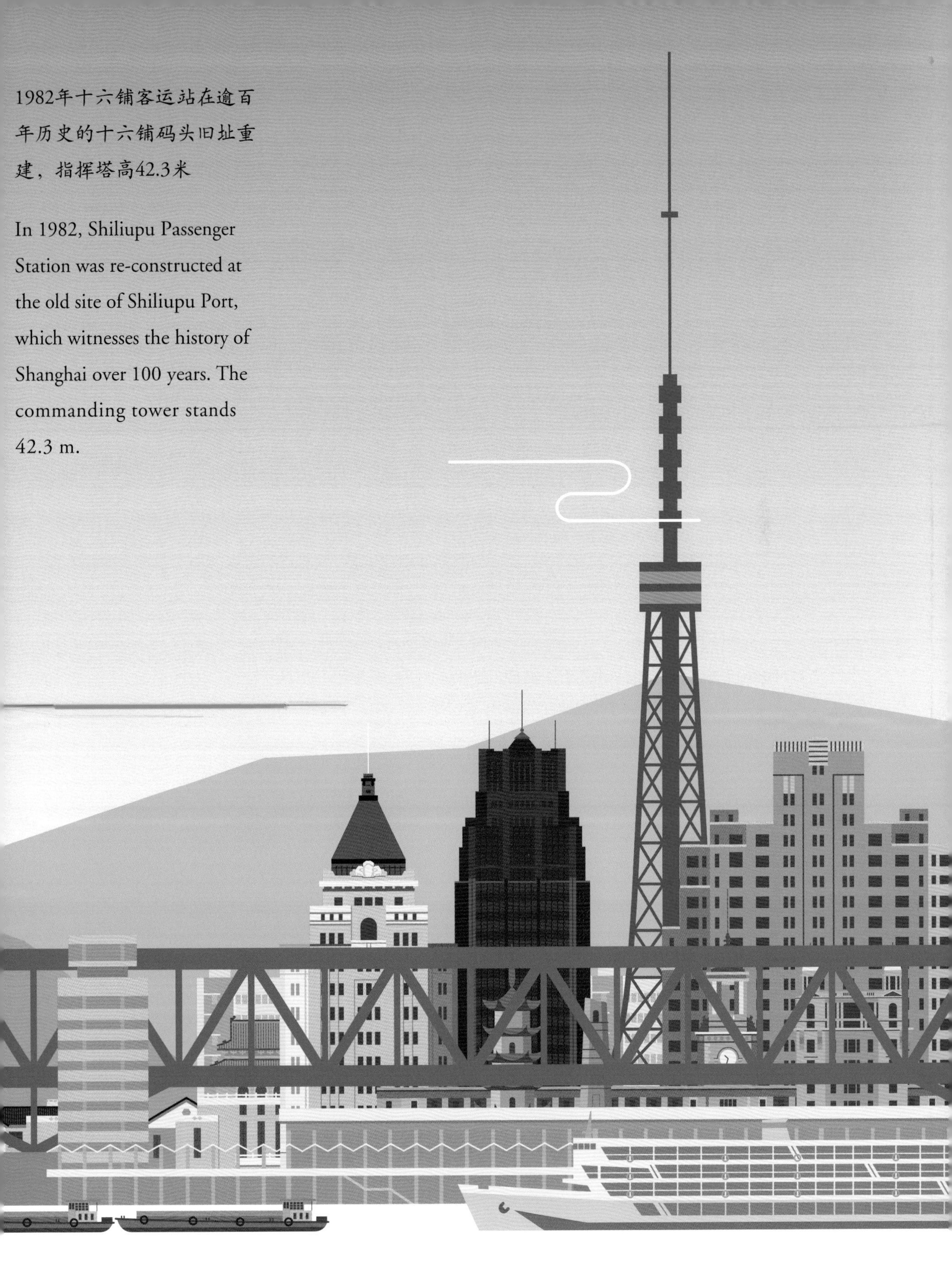

1982年十六铺客运站在逾百年历史的十六铺码头旧址重建，指挥塔高42.3米

In 1982, Shiliupu Passenger Station was re-constructed at the old site of Shiliupu Port, which witnesses the history of Shanghai over 100 years. The commanding tower stands 42.3 m.

第一幢高度超过国际饭店的建筑上海宾馆于 1983 年建成，高 90.5 米

Shanghai Hotel, the first building whose height surpassed Park Hotel, was completed in 1983 at the height of 90.5 m.

联谊大厦是上海第一幢高度超越百米的大楼，建于 1985 年，高 106.5 米

The Union Friendship Tower (106.5 m), built in 1985, is Shanghai's first building that reaches over 100 m.

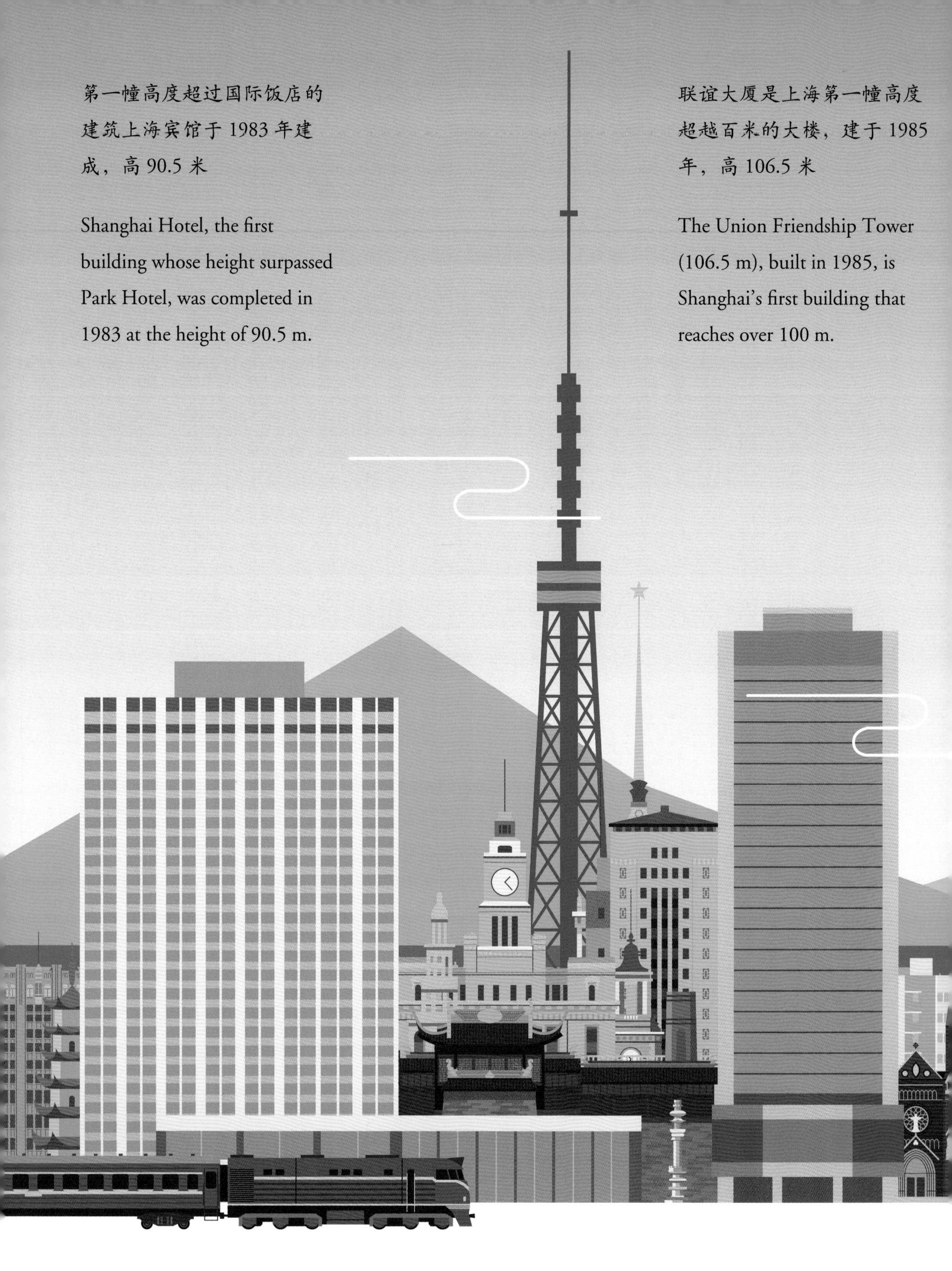

1988 年华东电力大厦建成，高 125.5 米，其独特的造型受人瞩目

In 1988, East China Electric Power Tower was completed with a height of 125.5 m. Its distinctive style was then the focus of the public.

上海第一家五星级酒店静安希尔顿酒店建于 1988 年，高 143.6 米

Shanghai's first five-star hotel, Jing'an Hilton, was completed in 1988 with a height of 143.6 m.

上海商城建于 1990 年，高 167 米

Shanghai Centre was built in 1990 with a height of 167 m.

杨浦大桥建于 1993 年，是当时世界上跨度最大的斜拉桥，大桥主塔高 208 米

The Yangpu Bridge, completed in 1993, was then the longest cable-stayed bridge in the world. The double-tower reaches a height of 208 m.

29. 静安希尔顿酒店
建于 1988 年，高 143.6 米
［香港］协建建筑师事务所设计
静安区华山路 250 号

30. 上海商城
建于 1990 年，高 167 米
［美］波特曼建筑设计事务所设计
静安区南京西路 1376 号（近西康路）

31. 杨浦大桥
建于 1993 年，大桥主塔高 208 米
上海市政工程设计研究院设计

32. 东方明珠广播电视塔
建于 1995 年，高 468 米
华东建筑设计研究院设计
浦东新区陆家嘴

33. 金茂大厦
建于 1998 年，高 420 米
［美］SOM 建筑设计公司设计
浦东新区世纪大道 88 号

34. 上海环球金融中心
建于 2008 年，高 492 米
［美］KPF 建筑师事务所设计
浦东新区世纪大道 100 号

35. 上海中心
建于 2015 年，高 632 米
［美］Gensler 建筑设计事务所设计
浦东新区银城中路 501 号

（信息截止日期：2015 年 8 月）

Index

1. Songjiang Dharani Column of the Tang Dynasty
Built in 859 (Tang Dynasty).
Remaining height: 9.3 m.
Full name: Usnisa Vijaya Dharani Sutra Column.
Location: In a school on East Zhongshan R., Songjiang District.
Important Cultural Heritage Construction under the Protection of Chinese Central Government.

2. The Qinglong Pagoda
First built during 821–824 (Tang Dynasty), rebuilt in Northern Song Dynasty. Remaining height: about 30 m.
Location: Qinglong Village, Baihe Town, Qingpu District.
Monument under the Protection of Shanghai Municipality.

3. The Longhua Pagoda
First erected (allegedly) in the period of the Three Kingdoms, rebuilt in the Northern Song Dynasty. Height: 40.64 m.
Location: 2872 Longhua Road, Xuhui Disctrict.
Important Cultural Heritage Construction under the Protection of Chinese Central Government.

4. Shanghai City Wall
Built in 1553 (Ming Dynasty).
Height: almost 8 m.
Location: 257 Dajing Road (near Renmin Road), Huangpu District.
Monument under the Protection of Shanghai Municipality.

5. Jixiang Li
Built before 1876. Height: 10–11 m.
Location: Lane 531, Middle Henan Road (near Ningbo Road), Huangpu Disctrict.
Shanghai Heritage Architecture.

6. Yangshupu Water Works
Established in 1883. Height of the remaining chimney: about 29 m.
Designed by J. W. Hart, contracted to S. C. Farnham & Co.
Location: 830 Yangshupu Road (near Xuchang Road), Yangpu Disctrict.
Important Cultural Heritage Construction under the Protection of Chinese Central Government.

7. The Gutzlaff Signal Tower
Re-built in 1908. Height: 49.8 m.
Location: No.2 East Zhongshan Road & East Yan'an Road, Huangpu District.
Important Cultural Heritage Construction under the Protection of Chinese Central Government (the Bund Architecture Complex).

8. St. Ignatius Cathedral
Built in 1910. Height: 56.6 m.
Designed by W. M. Dowdall.
Location: 158 Puxi Road (near North Caoxi Road), Xuhui District.
Important Cultural Heritage Construction under the Protection of Chinese Central Government.

9. Sincere Department Store
Now Shanghai Fashion Store and East Asia Hotel.
Built in 1917. Height: nearly 60 m.
Designed by Lester, Johnson & Morris, contracted to Gu Lan Kee Construction Company.
Location: 690 East Nanjing Road (near Middle Zhejiang Road), Huangpu District.
Monument under the Protection of Shanghai Municipality.

10. North China Daily News Building
Now AIA Building.
Built in 1923. Height: 40.2 m.
Designed by Lester, Johnson& Morriss, contracted to American Trading Company.
Location: 17 No.1 East Zhongshan Road (near Jiujiang Road), Huangpu District.
Important Cultural Heritage Construction under the Protection of Chinese Central Government (the Bund Architecture Complex).

11. The Great World Amusement Arcade
Now unoccupied.
Rebuilt in 1924. Height: 55.3 m.
Designed by Huinan Zhou, contracted to Senmao Construction Company.
Location: 1 South Xizang Road (near East Yan'an Road), Huangpu District.
Monument under the Protection of Shanghai Municipality.

12. The General Post Office of Shanghai
Built in 1924. Height: 51.16 m.
Designed by Stewardson & Spence, contracted to Ah Hung General Building Contractors.
Location: 276 North Suzhou Road (near North Sichuan Road), Hongkou District.
Important Cultural Heritage Construction under the Protection of Chinese Central Government.

13. The Customs House
Built in 1927. Height: 70 m.
Designed by Palmer & Turner, contracted to Sing King Kee General Building Contractors.
Location: 13 No.1 East Zhongshan Road (near Hankou Road), Huangpu District.
Important Cultural Heritage Construction under the Protection of Chinese Central Government (the Bund Architecture Complex).

14. The Sassoon House
Now the North Building of the Fairmont Peace Hotel.
Built in 1928. Height: 77 m.
Designed by Palmer & Turner, contracted to Sin Jin Kee Construction Company.
Location: 20 No.1 East Zhongshan

Road (near East Nanjing Road), Huangpu District.
Important Cultural Heritage Construction under the Protection of Chinese Central Government (the Bund Architecture Complex).

15. Park Hotel
Built in 1934. Height: 83.8 m.
Designed by László Hudec, contracted to Voh Kee Construction Company.
Location: 170 West Nanjing Road (near Huanghe Road), Huangpu District.
Important Cultural Heritage Construction under the Protection of Chinese Central Government.

16. The Broadway Mansions
Now Shanghai Mansion.
Built in 1934. Height: 76.7 m.
Designed by Palmer & Turner, contracted to Sin Jin Kee Construction Company.
Location: 20 North Suzhou Road (near Changzhi Road), Hongkou District.
Important Cultural Heritage Construction under the Protection of Chinese Central Government (the Bund Architecture Complex).

17. Shanghai Municipal Library
Now unoccupied.
Built in 1935. Height: 25 m.
Designed by the Greater Shanghai Plan Commision, administered by Dayou Dong, contracted to Chang Yue Tai Construction Company.
Location: 181 Heishan Road (near Zhengli Road), Yangpu District.
Shanghai Heritage Architecture.

18. The Sun Department Store
Now Shanghai No. 1 Department Store.
Built in 1936. Height: 42.3 m.
Designed by Kwan, Chu and Yang Architects, contracted to Voh Kee Construction Company.
Location: 830 East Nanjing Road (near Middle Xizang Road), HuangpuDistrict.
Shanghai Heritage Architecture.

19. Bank of China Building
Built in 1937. Height: 69 m.
Designed jointly by Luke Him Sau and Palmer & Turner, contracted to Doe Kwei Kee General Building Contractor.
Location: 23 No.1 East Zhongshan Road (near Dianchi Road), Huangpu District.
Important Cultural Heritage Construction under the Protection of Chinese Central Government (the Bund Architecture Complex).

20. Caoyang No.1 New Estate
First built in 1952. Height: 11–12 m.
Designed by Shanghai Municipal Architectural Design Company.
Location: Lanxi Road (near Huaxi Road), Putuo District.

21. Sino-Soviet Friendship Building
Now Shanghai Exhibition Center.
Built in 1955. Height: 110.4 m.
Designed jointly by Soviet experts and East China Architectural Design Institute.
Location: 1000 Middle Yan'an Road (near Weihai Road), Jing'an District.

22. Shanghai Television Tower
Completed in 1974. Height: 210 m.
Designed by Tongji University.
Demolished in 1998.
Location: Qinghai Road (near West Nanjing Road), Jing'an District.

23. Xuhui New Estate
Constructed in 1976. Height: 38.75–51.55 m. 9 Buildings.
Designed by Shanghai Municipal Institute of Civil Architectural Design.
Location: 750–1000 North Caoxi Road (near Yude Road), Xuhui District.
Shanghai Heritage Architecture.

24. Huangpujiang Bridge
Now Songpu Bridge.
Completed in 1976. Height: over 50 m.
Designed by the Bridge Engineering Bureau, Ministry of Transport.
Location: Cheting Highway (near Chedun Town), Songjiang District.

25. Shiliupu Passenger Station
Now transformed into Riverside Walking Path.
Built in 1982. Tower Height: 42.3 m.
Designed by East China Architectural Design Institute.
Demolished in 2004.
Location: Along No.2 East Zhongshan Road, Huangpu District.

26. Shanghai Hotel
Built in 1983. Height: 90.5 m.
Designed by Shanghai Municipal Institute of Civil Architectural Design.
Location: 505 North Wulumuqi Road (near Huashan Road), Xuhui District.

27. The Union Friendship Tower
Built in 1985. Height: 106.5 m.
Designed by East China Architectural Design Institute.
Location: 100 East Yan'an Road (near Middle Sichuan Road), Huangpu District.

28. East China Electric Power Tower
Built in 1988. Height: 125.5 m.
Designed by East China Architectural Design Institute.
Location: 201 East Nanjing Road (near Middle Henan Road), Huangpu District.

29. Jing'an Hilton Hotel
Completed in 1988. Height: 143.6 m.
Designed by Hong Kong AP Architects Ltd.
Location: 250 Huashan Road, Jing'an District.

30. Shanghai Centre
Built in 1990. Height: 167 m.
Designed by John Portman & Associates, Inc.
Location: 1376 West Nanjing Road (near Xikang Road), Jing'an District.

31. Yangpu Bridge
Completed in 1993. Tower Height: 208 m.
Designed by Shanghai Municipal Engineering Design Institute.

32. The Oriental Pearl Radio & TV Tower
Completed in 1995. Height: 468 m.
Designed by East China Architectural Design & Research Institute.
Location: Lujiazui, Pudong New Area.

33. Jin Mao Tower
Completed in 1998. Height: 420 m.
Designed by Skidmore, Owings & Merrill LLP (SOM).
Location: 88 Century Avenue, Pudong New Area.

34. Shanghai World Financial Center
Completed in 2008. Height: 492 m.
Designed by Kohn Pedersen Fox (KPF).
Location: 100 Century Avenue, Pudong New Area.

35. Shanghai Tower
Completed in 2015. Height: 632 m.
Designed by M. Arthur Gensler Jr. & Associates, Inc. (Gensler).
Location: 501 Middle Yincheng Road, Pudong New Area.

(Last update: Aug 2015)

本社出版的延伸读物及“城市行走”书系
Book Recommendation & *CityWalk* Series

《上海百年建筑史：1840—1949》，伍江著

A History of Shanghai Architecture 1840-1949　Wu Jiang

ISBN 978-7-5608-3895-3
2008
¥: 50.00

《上海外滩建筑地图》
乔争月、张雪飞著

Shanghai Bund Architecture
Michelle Qiao, Zhang Xuefei

ISBN 978-7-5608-5867-8
2015
¥: 48.00

《上海教堂建筑地图》
周进著

Shanghai Church
Zhou Jin

ISBN 978-7-5608-5675-9
2014
¥: 58.00

《上海邬达克建筑地图》
华霞虹、乔争月等著

Shanghai Hudec Architecture
Hua Xiahong, Michelle Qiao

ISBN 978-7-5608-5061-0
2013
¥: 56.00

《上海里弄文化地图：石库门》
姜庆共、席闻雷著

Shanghai Shikumen
Jiang Qinggong, Xi Wenlei

ISBN 978-7-5608-4791-7
2012
¥: 42.00

《上海城市雕塑地图》
屠娟编著，栾建红译

Shanghai City Sculpture Map
Tu Juan, Luan Jianhong

ISBN 978-7-5765-0216-9
2022
¥: 45.00

《上海杂货铺》
周祺著

Shanghai Housewares
Zhou Qi

ISBN 978-7-5608-5240-9
2013
¥: 48.00

《上海张爱玲文学地图》
淳子、王桢栋、冯宏、营光学社著

Eileen Chang's Shanghai
Chun Zi, Wang Zhendong, Feng Hong, Highlight Studio

ISBN 978-7-5608-7573-6
2019
¥: 62.00

图书在版编目（C I P）数据

上海高度 = Shanghai Height : 汉、英 / 姜庆共著
. -- 上海 : 同济大学出版社, 2015.8
（城市行走 / 江岱, 姜庆共主编）
ISBN 978-7-5608-5937-8

Ⅰ. ①上… Ⅱ. ①姜… Ⅲ. ①建筑物－介绍－上海市－汉、英 Ⅳ. ①TU-862

中国版本图书馆CIP数据核字(2015)第181107号

上海高度

姜庆共 著
出品人：支文军
责任编辑：江岱
助理编辑：罗璇
责任校对：徐春莲
出版发行：同济大学出版社 www.tongjipress.com.cn
地　　址：上海市四平路1239号 邮编：200092
电　　话：021—65985622
经　　销：全国新华书店
印　　刷：上海雅昌艺术印刷有限公司
开　　本：787mm × 1092mm　1/16
印　　张：2
字　　数：49 000
版　　次：2015年8月第1版
印　　次：2024年5月第2次印刷
书　　号：ISBN 978-7-5608-5937-8
定　　价：36.00元